Pollinators

MOTHS

DiscoverRoo
An Imprint of Pop!
popbooksonline.com

Emma Bassier

abdobooks.com

Published by Pop!, a division of ABDO, PO Box 398166, Minneapolis, Minnesota 55439. Copyright © 2020 by POP, LLC. International copyrights reserved in all countries. No part of this book may be reproduced in any form without written permission from the publisher. Pop!™ is a trademark and logo of POP, LLC.

Printed in the United States of America, North Mankato, Minnesota.

102019
012020

THIS BOOK CONTAINS RECYCLED MATERIALS

Cover Photo: Frank Hecker/Alamy
Interior Photos: Frank Hecker/Alamy, 1; Gregory G. Dimijian, M.D./Science Source, 5; A. B. Joyce/Science Source, 6; iStockphoto, 7 (flowers), 7 (moths), 11, 16 (bottom), 17 (top), 17 (bottom), 19, 20, 21, 23, 27, 30; Pam Collins/Science Source, 8–9; Ted Kinsman/Science Source, 12; Shutterstock Images, 13, 26; The Natural History Museum, London/Science Source, 14; Nature's Images/Science Source, 15, 31; Kenneth M. Highfill/Science Source, 16 (top); Merlin Tuttle/Science Source, 22; Stuart Wilson/Science Source, 25; NASA, 28–29

Editor: Connor Stratton
Series Designer: Jake Slavik
Library of Congress Control Number: 2019942475
Publisher's Cataloging-in-Publication Data

Names: Bassier, Emma, author.
Title: Moths / by Emma Bassier
Description: Minneapolis, Minnesota : Pop!, 2020 | Series: Pollinators | Includes online resources and index.
Identifiers: ISBN 9781532165986 (lib. bdg.) | ISBN 9781532167300 (ebook)
Subjects: LCSH: Pollinators--Juvenile literature. | Moths--Juvenile literature. | Moths--Behavior--Juvenile literature. | Pollination by insects--Juvenile literature. | Insects--Juvenile literature.
Classification: DDC 595.78--dc23

WELCOME TO
DiscoverRoo!

Pop open this book and you'll find QR codes loaded with information, so you can learn even more!

Scan this code* and others like it while you read, or visit the website below to make this book pop!

popbooksonline.com/moths

*Scanning QR codes requires a web-enabled smart device with a QR code reader app and a camera.

TABLE OF
CONTENTS

CHAPTER 1
NIGHT FLYERS

The sun sets. A furry insect flies across the orange sky. It is a moth. The moth lands on a yellow flower. Its hairy body gets covered in **pollen**. Then the moth flies to another plant.

WATCH A VIDEO HERE!

A moth visits a senita cactus flower that is filled with pollen.

Moths visit flowering plants for a couple of reasons. One reason is to eat. Moths sip **nectar** from inside the flowers. Moths also sometimes land on plants to lay eggs. When the moths fly to other plants, they spread pollen.

Moths lay eggs on the leaves of a flowering plant.

MOTH POLLINATION

A moth lands on a flower. Pollen gets on the moth's body. The moth flies to another flower. Some of the first flower's pollen falls into the second flower.

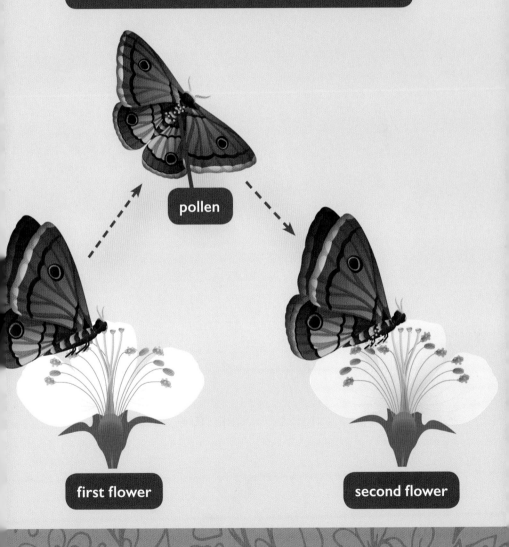

pollen

first flower

second flower

Pollinators such as moths carry

pollen from flower to flower. Then plants

use that pollen to make new seeds.

Night-blooming jasmine is one type of flower that moths help pollinate.

Without moths to carry their pollen,

some plants might not survive.

DID YOU KNOW? Moths can travel long distances. One type of moth can fly for 100 miles (160 km) at one time.

CHAPTER 2
FEATHERY FEELERS

More than 160,000 types of moths exist. All moths have antennae. These feelers stick out from the moth's head. They are feathery and thin. Moths use their antennae to smell.

LEARN MORE HERE!

antennae

Cecropia moths have large, feathery antennae.

DID YOU KNOW? One type of male moth can smell a female moth 7 miles (11 km) away.

A close-up picture shows the scales of a luna moth's wing.

Many moths have dull wing colors.

They may be tan, brown, or gray. Moths

have many tiny scales on their wings.

The scales can look like dust or powder.

A sunset moth's wings can show nearly all colors in the rainbow.

To sip **nectar**, many moths use

a feeding tube known as a proboscis.

A moth stick its proboscis into a flower.

Then it sucks up the flower's nectar.

proboscis

One type of sphinx moth has a proboscis
that is 1 foot (0.3 m) long.

A yucca moth visits a yucca flower in Texas.

Some moths do not have mouthparts. These moths have short lives. As a result, they do not need to eat as adults.

YUCCA PLANTS AND MOTHS

Yucca moths do not have proboscises. Instead, each moth has two short tentacles. It uses these mouthparts to collect **pollen** from yucca plants. The moth forms a ball of pollen. Then it flies to another yucca plant. The moth drops the pollen and lays eggs there. The plant uses the pollen to grow fruit and seeds. The fruit feeds the young moths. In this way, the moth and plant help one another survive.

LIFE CYCLE OF A MOTH

An adult female moth lays eggs.

Caterpillars hatch from the eggs. Some moth eggs hatch after a few days. Others take many months.

Caterpillars spend almost all their time eating. They grow bigger and bigger.

Caterpillars make protective coverings known as cocoons. Their bodies change completely while inside.

Adult moths break out of their cocoons.

Some moths only live a few days. Others live several months.

CHAPTER 3
BLENDING IN

Moths live all around the world. To survive, they **adapt** to where they live. For example, many moths live in forests. These moths' wings often look

COMPLETE AN ACTIVITY HERE!

Moths often look for the place on a tree that hides them best.

like bark and leaves. This **camouflage**

helps them hide from other animals.

Some moths **mimic** other animals in their **habitats**. For instance, moths often have eyespots on their wings. These spots look like the eyes of larger animals. They help scare off **predators**.

Some moths have eyespots that look like an owl's eyes.

eyespot

Hummingbird hawk moths mimic hummingbirds. These moths even flap their wings like hummingbirds.

DID YOU KNOW?

Some moths look similar to wasps or spiders. Others look like bird poop.

Many bats use echolocation to find food, such as moths. They squeak and listen for the echoes.

In addition, most moths are active at night. This helps them avoid being seen and eaten by predators. However, many bats eat moths. Bats hunt at night.

They often use sound to find food.

In response, some moths make their

own sounds. These squeaks confuse bats

so the moths can escape.

*The dogbane tiger moth makes sounds
to confuse attacking bats.*

CHAPTER 4
PROTECTING MOTHS

Moths face a variety of threats. One

big danger is human-made light.

At night, moths can often fly safely.

But streetlights show where moths are.

Bats can easily see and eat them.

LEARN MORE HERE!

Many moths, including certain sphinx moths, are at risk of dying out completely.

In addition, moths use moonlight to find their way. Other kinds of lights can confuse them. The moths fly to those lights instead of flowers. These moths spend less time pollinating. As a result, certain plants make fewer seeds.

Moths do not eat when they are near streetlights. A single light can cause many moths to starve every night.

Some moths, such as the ermine moth, have adapted to avoid city lights.

Moths help pollinate a large number of crops, including peas and soybeans.

However, cities can turn off their

streetlights at night. That way, moths do

not become confused by human-made

A picture taken from space shows human-made lights on Earth.

light. This action helps moths survive and

pollinate. By protecting moths, people

can care for life on Earth.

MAKING CONNECTIONS

TEXT-TO-SELF

Have you seen moths before? If so, what kinds?
If not, where might you find them?

TEXT-TO-TEXT

Have you read books about other insects?
What do they have in common with moths?
How are they different?

TEXT-TO-WORLD

Moths use their wing colors and patterns
to blend in. What is another way animals can
stay safe from predators?

GLOSSARY

adapt – to develop traits that make it easier to survive in an environment.

camouflage – a pattern, color, or shape that helps an animal blend in with its surroundings.

habitat – the area where an animal normally lives.

mimic – to copy how something else looks or behaves.

nectar – a sweet, sugary liquid made by a plant.

pollen – fine, dust-like stuff that flowers create and use to reproduce.

predator – an animal that hunts other animals for food.

INDEX

 ONLINE RESOURCES

popbooksonline.com

Scan this code* and others like it while you read, or visit the website below to make this book pop!

popbooksonline.com/moths

*Scanning QR codes requires a web-enabled smart device with a QR code reader app and a camera.